Where Neuroscience Meets Art

Frank Stringfellow BS, MA, Ph. D.

To order additional copies of this book, contact:
Xlibris
844-714-8691
www.Xlibris.com
Orders@Xlibris.com

ISBN: 978-1-6698-7876-6 (sc)
ISBN: 978-1-6698-7875-9 (e)

Library of Congress Control Number: 2023910041

Print information available on the last page

Rev. date: 06/01/2023

Where Neuroscience Meets Art

by

Frank Stringfellow BS, MA, Ph. D.

CONTENTS

Where Neuroscience Meets Art
Pattern Recognition and Mirror Neurons,
Implications for Mapping the Human Brain
from
Collected Works of Frank Stringfellow
(Effects of Aging on Creativity and Expression)
by

Frank Stringfellow BS, MA, Ph. D.

UPGRADES OF:
TXu1-043-426 (7/19/01)
SRu 561-934 (4/26/04)
SRu 866-640 (8/30/07)
SRu 1-046-548 (6/3/2011)
SRu 1-293-075 (3/16/2016)
TXu 2-267-678 (6/18/2020)
Txu 2-320-553 (1/18/2022)

PATTERN RECOGNITION AND MIRROR NEURONS,
IMPLICATIONS FOR MAPPING THE HUMAN BRAIN

Where Neuroscience Meets Art

"THE DRAW"

What is the "Draw" and how has it gotten me to this point in time? (1/9/2022).

It is you being pulled to do something. Find pleasure in something, want that, and make it happen. To reach out beyond the self.

I have been given many gifts at birth. A good brain, mind, and personality which translates into actions done easily like, drawing, painting, woodburning, writing, analyzing, storytelling, critical thinking, and science. I feel the obligation to make use of those gifts to do something positive with them.

There are certain *emotions* that I don't seem to have or they are weak. They are fear based and distractive such as envy, jealousy, and greed and they are negative. These emotions are used a lot in advertisements to influence people to do something someone wants them to do. I have never sought *power* or influence and that is readily shown in how I have lived. People seek power out of fear-based emotions. They believe that it gives them control over their lives. Nothing could be further from the truth because they are always guarding their turf or defending their position. Looking over their shoulder! My attitude has always been look ahead, travel light, and leave your fears behind you and that is how I have lived my life. Some of "looking ahead" is: To see what is happening on the other side of the hill, first, you have to climb to the top of the hill.

It may be that the lack of those negative emotions which can distract or hold you back enable positive emotions/actions to emerge and be expressed as my gifts. I am then compelled to exercise those gifts whether there is a benefit to me or not. It usually comes in the form of some stimulus which produces memories from past experiences.

Why am I doing this paper? It popped-up into my head as I was out exercising this morning (11/4/2021). I came home and I felt the need to write since "The Draw" is such an integral part of the exercise of all of my gifts and in this case direction.

I have to express a gift or it won't go away! I always get that warm "fuzzy" feeling (neurotransmitters) of satisfaction and accomplishment after I do that. That is my reward (my happy place) much as I describe when someone "lights up" over the expression of one of my gifts.

TACTICS AND STRATEGY (1/1/2020)

How have I done all of this? I always said that there never was a plan. I read an article that explained about tactics and strategy and I realized that I didn't know the difference between them. I looked them up in the dictionary and apparently tactics is the short-term plan to reach a goal and strategy is a long-term plan to reach a goal. I didn't need a plan because it was a part of me. This really was the *strategy* (1) I used for long term planning which is really no plan at all. It was all tactical planning.

How and why have I gotten to where I am since there was no plan? I have been blessed with many gifts from birth and I simply used my gifts and followed my interests, taking one step at a time.

I worked on the *Collected Works of Frank Stringfellow* in my spare time. It was my passion and fun to me. It was never work. It was to preserve family history and my science and artworks since I am a major part of family history. I was just having fun!

Where this will go is dependent on outside resources and connection with the right people and cooperation from us.

The above is a good example of the journey and my brain research.

PREVIEW

Refer to the schematic.

Think of this as an overview (flyover).

We are going to see what happens in the human brain at that point of connection between the artist and the observer using artwork as data. The artwork was done by me over several years (1980 at least > 42 years). I said in 2002 after I retired that I was going to do brain research: *The Effects of Aging on Creativity and Expression.*

Let's look at the Artist:

We all have life's experiences and they are different for each person. These experiences are taken into the brain by way of the senses: sight, hearing, taste, smell, and touch etc. These experiences are stored in the brain as memories/thoughts. Stimuli presented through the senses cause the mind to see similarities (pattern recognition) between the stimuli and the stored memories/thoughts to recall the memories/thoughts from the brain and with the mind/personality (lights up) acts out to produce an artwork. Another way of looking at this is that the mind is an expression of the brain and the personality is expressed by the mind. The person loves that, finds pleasure in that, has to have that, and acts out producing the artwork. He makes it happen.

The artist has now produced an artwork. He found pleasure in the memory that caused the mind to recall that and expressed it through the personality.

The brain/mind/personality complex are self-awareness. Acting out to produce a tangible creation is consciousness. It is an action outside of the self.

The artwork can be a poem, painting or even a concept that an engineer designs to solve a problem.

The artist has now produced an artwork. He found pleasure in the memory that caused the mind to recall that and expressed it through the personality.

The Observer:

An observer comes along and he loves that artwork (lights up), finds pleasure in it, wants that, and makes it happen, He buys the artwork. There is then some sort of empathy going on between the artist and observer. That empathy I call the "Draw" or "Pull" common to both the artist and the observer.

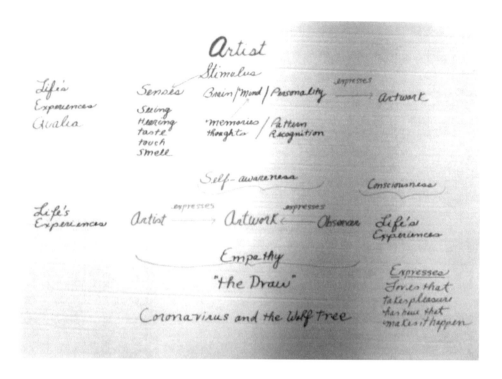

CORONAVIRUS AND THE WOLF TREE

I will use the *Coronavirus and the Wolf Tree* to illustrate it with several other examples if time permits.

Example: CORONAVIRUS AND THE WOLF TREE. See: # 351 and # 354 in Icon # 13.

I would never have thought about the wolf tree until the coronavirus pandemic occurred. There are a lot of similarities and free associations with the wolf tree. This is good brain research where the similarities between the virus and the tree led to the creation of the artwork **(pattern recognition = memories)**. The coronavirus is **scary** and hand-washing, distancing, and masking created a sense of **isolation** and **loneliness** making people **stir crazy**. I have attempted to make artworks of the wolf tree which reflects this sense of isolation and loneliness. The **World on Fire** came from the program on *MPT*! I borrowed it. I thought it was appropriate.

I have actually seen a wolf tree. We were out on a field trip in 1967 in South Carolina when I was a graduate student and Dr. Wade Batson pointed it out. They usually are white oaks in the strictest sense. Wolf trees are isolated from the surrounding trees because they were there long before the other trees germinated. They are **bigger** than the surrounding trees for the same reason. They have **branched** many times and are big and **spreading**. The **multi-branching** is caused by damage to the apical meristems as they grow. The appearance is scary looking and it is isolated from the other trees. It is spreading in all directions and trails off into infinity. I would like to see one at night! See: Icon 13.

The wolf tree(virus) is vulnerable as shown by the burning limb. If the virus gets into a compatible host it will survive, but if it gets into a highly susceptible host it will die with the host so it is in its best interest to not kill its host. The coronavirus is ubiquitous throughout the human population and emerges periodically as highly virulent strains: SARS, MERS, and COVID-19 etc.

Here are **the three fingers of death**. The death rates have been less than those caused by influenza.

Here is a branch that looks like a **headless human** signifying the impact of the virus on the human population (**panic**).

There are at least 10 **strongpoints** (associations) which resonated with me. Others are precautions (**masking, handwashing, and distancing**).

The end result was that *when the coronavirus hit and spread with its attendant associations, it brought up the memory of the wolf tree and they came together and I produced this painting. I had to have that!* It is a good example of long-term memory.

The physical basis for the coronavirus and the wolf tree is in the limbic system which comprises cortical and subcortical areas of the brain.

Short term memory was consolidated into **long term memory** soon after I saw the wolf tree. It was explicit, episodic autobiographical memory and part of my life's experiences.

I could see meaningful connections (pattern recognition) between unrelated things. Such as, spreading, isolation, loneliness, multi-branching, bigger than surrounding trees, death, the effects on mankind (headless branch fleeing), and fire and secondarily masking, handwashing, and distancing. It is called template matching of news (stimuli) from all media (TV, radio, newspapers).

Apophenia is a term Dr. Schultz uses to describe those who can see patterns where there are none to see by others, to see more than is there. (Schultz, William Todd. *The Mind of the Artist. Personality and the Drive to Create*. Oxford. 2022. 200 pp.) many of the strongpoints for both of these paintings came from the media (radio, newspapers, TV).

How then does pattern recognition come into this?

Definition of pattern recognition (Wikipedia): The cognitive process (producing an artwork) that matches information from a stimulus (coronavirus and resulting patterns of...) retrieved from memory (wolf tree).

107. *# 351. Coronavirus and the Wolf Tree (The World on Fire). (6/12/2020).* Acrylic on canvas. 36 X 48 inches. See # 333.

351 led to # 354, *Burn, Baby, Burn.*

BURN, BABY, BURN

This painting depicts riots happening in Baltimore, Seattle and around the country. I used painting # 31 *Burn, Baby, Burn* as a guide.

It depicts rioting, looting, and burning of property. I was caught in a riot in Washington, D.C. around 1970. I had just started work at the USDA and we were invited to the Smithsonian to see some of their work. We were told to abandon the building for a riot was headed our way. We did and went out the south side of town. People were milling around us but they were not violent and didn't try to do anything destructive. Later on, there was looting, burning, and destruction of property. Mayor Washington stood tall!

How did I arrive at this painting? I have always had a fascination with fire (# 16, # 95, # 334, # 335, # 350, # 351).

My brother and I used to get cardboard boxes and put matches on them and light them off. We didn't grow up to become arsonists. Apparently, the people depicted in this painting did. The news stimulated those old memories through my senses and I loved that, had to have that, and made it happen! # 353. Those memories were stored in my brain all those years, restimulated through my visual and auditory senses, were recalled by my mind and was expressed by my personality as this painting. This is an example of pattern recognition frog hunting style. The vanishing point for this painting is in the center street where a person is burning. This represents those killed in the fires!

I wrote a poem entitled *Burn, Baby, Burn* and penned it on the back of this painting.

BURN, BABY, BURN

Free will passed away today.
    Smothered by a blanket
        Of "political correctness."
Al didn't do it.
It happened through controlling
    Talk and actions, or
        Shouting you down!

My opinions based on facts
    Will not "Trump" your opinions
        Based on feelings.
Hatred for America is American fire
    Smoldering globally, yet
        Everyone wants to come here
For easy pickings?
    Or
The American "Dream!"
One door closes. Another door opens!

(4/23/2021)

110. *# 354. Burn, Baby, Burn.* (5/23/2021). Mixed Media (acrylic, ink, oil, woodburning, spray). 8 X 6 feet.

Apparently, these memories are stored in the amygdala. I matched what I was experiencing with what I experienced, liked it, wanted it, and acted out and painted it just as I had with # 351.

I liked this poem, these paintings because they describe how I felt about doing them and the satisfaction I have when I see them again.

The feelings I get when I like what I see, act out and do them, am satisfied in reading and doing them again are a result of neurotransmitters (dopamine and serotonin) operating. Apparently, these involve the reward system in the brain.

The anatomical basis described in this paper came from Wikipedia (**W**) where the original work is explained and referenced. I did not do the original works but I used them as a source.

Enter terms like long term memory, amygdala, limbic system, rewards system into Wikipedia and it will give greater detail. I have shown the way by keeping things simple.

PERSONALITY, SELF-AWARENESS, AND CONSCIOUSNESS

Personality and Self-Awareness: The end result of this investigation is that personality influences the search for Self-awareness/Consciousness (S/C). Personality is influenced by genetics, environment, and culture. It is the mind expressing itself. That is why there is no one path for everyone. That is why I arrived at the conclusion I did. Any other personality might arrive at a different conclusion. Personality may fit a type, but it is evident that there are infinite degrees of personalities within a type and the same applies to S/C. I fit the type of conscientiousness/openness in the Five Factor Model.

I began this study in 2007 because it was obvious that self-awareness and consciousness had become confusing and so complex so as to be incomprehensible. *(Blackmore, Susan, and Troscianko, Emily T. Consciousness an Introduction. 3rd Edition. Routledge, London and New York. 617 pp. 2018).* This has resulted in a simplification process in which self-awareness is basically recognizing one's individuality and consciousness is thinking outside of the self and expressing the self in tangible creations. They can range from works of art such as paintings, wood burnings, poems, models, sculptures or even designs and equipment done by an engineer to solve a problem. But only, if he is "making his baby."

AWARENESS, CONSCIOUSNESS, AND ART (12/29/10):

How does the brain create the mind? The conscious mind is holistic. It is the sum of all of the brain's parts above the autonomic level. The brain through its various senses can detect its environment and change internally and externally accordingly (adapt). I can talk about nucleic acids, neurons, neurotransmitters, the multitude of cells and pathways and parts of the brain used to provide the final perception; but in the end the nervous system is basically an electrical system. It is always on and streaming (processing) conscious and unconscious information. It is magnetic and invisible but can be detected or measured indirectly as action potentials and with fMRI (blood flow). The brain is the site of all memories and thoughts and pathways which I propose could be electromagnetic (saltatory conduction-Nodes of Ranvier). Magnetism can fluctuate infinitely and represent any thought or memory. It is integral that recall of some memories and thoughts are instantaneous (pop-ups?). You can measure electrochemical action potentials or changes in the brain with fMRI or change the brain with TMS (transcranial magnetic stimulation). If you flat line you are brain dead and there are no more action potentials to record. Another way to visualize the mind is by producing tangible products such as artworks and interpret them as I have done in this *Works*.

I arrived at my ideas on self-awareness, consciousness, and art without any formal definitions of the terms, only my own search. This means there were no restrictions on the search. I plowed my own path and went back to basics (observation) since I am a scientist. I believe Dr. Gazzaniga got it right when he says that consciousness is an instinct (Gazzaniga, Michael S. *The Consciousness Instinct Unraveling the Mystery of How the Brain Makes the Mind*. Farrar, Straus and Giroux, NY, 2018 pp. 274.) However, instinct is a blanket term which does not give the details (particulars) of it. I have taken a similar approach in not spelling things out because the brain is so complex it would be difficult to pinpoint any one part as a source of memory. I have tried to capture its *essence*.

When I reached the end-point(s), I realized that there were many paths that I could have taken and which could have led to different conclusions. Example, you don't believe in God. That path might lead you to equate self-awareness and consciousness as the same thing since it would lead to self-absorption and the inability to look *beyond the self*.

I went back and found that the definitions could be interpreted differently by each individual. That accounts for the different conclusions that people have reached about self-awareness and consciousness and how they use these terms.

I picked out simple working definitions which *loosely* fit in with what I arrived at.

Self-awareness: An awareness of one's own personality or individuality.

Consciousness: The *upper level* of mental life of which the person is aware as contrasted with the unconscious processes. Reaching out beyond the self appears to be important.

Conscious: Awake (aware) or not awake (unaware).

Conscience: Sense of morality or doing what is right.

The above definitions are from the Merriam Webster dictionary as well as my own definitions.

I like asymmetry and simplicity and you will find it quite often in my paintings. I say that a painting is done when I see it as right (Qualia). I simply visualize the final painting and I go there. I don't believe that there is much outright analysis going on. I maintain that you could do all of the analysis and design and come up with a painting that is completely sterile.

So, what is beauty? The standard definition is that it is in the eye of the beholder. Maybe other questions are: What is chemistry, friendship, love?

WHAT IS BEAUTY?

NEURASTHETICS

The Mind, Brain Institute (Johns Hopkins University) Baltimore, MD in conjunction with the Walters Art Museum was giving a show called *What is Beauty* in 2010?

Our paths crossed when I was looking for new art shows in town. I saw *Neurasthenics* and I delved into it. It said that artists **are in their own way** neuroscientists. I went further into it and saw that we were doing similar work but in different ways. Their work was more objective (by putting a number on it) and mine was/is more subjective (by generating data as artworks and interpreting them with what I have read about the brain in the literature). What they did and my opinion about it is in the *Collected Works of Frank Stringfellow (pages 251-252)*. I will write about only what I am doing for brevity's sake.

Basically, what I am saying in my *Collected Works* is that the basis for all of the exercise of my gifts given to me from birth has yielded artworks: [paintings, wood burnings, poems, stories (written and spoken), essays etc.] all are based in memories and that the memories are based in life's experiences as perceived through the senses. That the memories are stored in the nucleic acids of the neurons and their inter-connections. That likely short-term memory is based in RNA because it has a short half-life and long-term memory is based in DNA. I don't know if or how proteins fit into this. That one-day individual memory may be assessed by the development of some process "out there." However, right now only one means exists to recall an individual's memories and that means is called the brain.

MY EXPERIMENTS

1. My process is getting the artworks to the point where I say that it looks right to me (Qualia?). That means that I have drawn on all of my life's experiences to get that painting where I want it. The end point is more important to me than the process of getting there. I usually know how I am going to get there if I can picture in my mind how I want it to be. Most of my paintings begin with a horizontal line or I start with the background and move forward. Nearly all of my artworks have a story attached to them. I believe that what I paint tells a story about me. The story behind the work of art embellishes it!

2. How do I measure response? By observation. Observation is a fundamental tool in science. It can be direct (artworks) or indirect (fMRI). If I take someone on a tour of my artworks invariably at some point, they will shine (light up). That is my reward. That means to me that I have connected or communicated with them in some way. Which ones do you like the most? Which ones do you like the least? Why? The painting can be beautiful (# 49-*Sunflowers*) or it can be a good painting but I wouldn't want it hanging in my living room. Example: (# 65-*Temple of the Warriors*). The observer lit up (shined), was neutral, or was repulsed by the painting because they saw something in the painting that moved them. It is the recall of memories of past experiences that caused that painting to appeal or not appeal to the observer. So, the observer and I have connected in some way (communicated).

How do I put a number on the creativity and expression? I have put numbers on some things. The number of expressions of each gift over time. Example: the number of artworks, the number of poems over time. Dates! It is left up to others to say about the quality of the creativity and its expression. That is subjective.

I can go through all of the design elements of making a good painting which produces a harmony that is pleasing to the brain (mind?) and which touches the very soul of a person's life experiences in such a way that they have to have (love) that. I call it "The Pull" (the Draw) (# 158 *The Lantern*). The painting can be either beautiful, neutral or repulsive to the observer based on their life's experiences which produce memories. You hear stories of the fear of walking up to a blank canvas not knowing what to paint (pushing). I have never had that problem because I get an idea in my head and I have to express it in artwork before it will go away (pulled). The painting has to looks right to me. (# 150 *Niagara Falls 2*).

One day I was coming home from my sister-in-law's and a poem popped into my head as I turned the corner. I parked, walked across the street to home and sat down and wrote a rough draft of it out on paper. The poem was *Tickle Me with Feathers*. I loved it. I had to have it!

"THE DRAW"

Is the draw (a reaching out beyond the self) and is this *Collected Works of Frank Stringfellow* (the tangible product) an expression of consciousness? Yes, because it reaches out beyond just me, to make a contribution to human knowledge, the world of art, and to preserve my works and family history.

What is that point of *connection* where the artwork that the artist produces elicits a response in the observer? Does it have a location structurally in the brain? Does it have survival value? Behaviorally, a person is drawn to something if they find pleasure in it (reward). They avoid something if they find pain in it or they are neutral on it at best. This is human nature. How far down the evolutionary chain this goes I do not know? It goes, at least, to the level of cephalopods (squid and octopus) and insects (crickets and cockroaches).

I have had experiences during my life time. These experiences have yielded memories in my brain. I conjure up these memories to produce works of art (poems, stories, paintings, carvings, woodburning etc.) after I have incubated them and then I act out on them. I visualize the end point and I go there either by constructing the fine elements to make a good painting (design), or by just going there, or by a combination of both of these devices. Nearly every work of art that I do has a story with it from my experiences.

The observer either responds to the artwork or they don't (neutral). The observer has their own set of experiences during their life time which produced memories which may not be similar to mine. The response of the observer is dependent upon their incubated memories. Let's say that the response is positive (drawn). I have communicated (connected) with the observer in some way. What is that point of connection? The connection between the artist and the observer I believe it is some form of empathy. Perhaps the memories of the observer are similar enough to that of the artist that they set off mirror neurons that makes the observer want (love) that. That is one explanation I have been able to come up with. What are mirror neurons? These are neurons that will light up with fMRI in the same part of the brain of a person eating a hot dog and when a person sees someone eating a hot dog. I wish I had a hot dog! Is this a form of brain mapping? If long term memory is in the cerebral cortex and other areas, can you create a map to locate where it is? See: *Experimental Design for Brain Mapping using Mirror Neurons and Pattern Recognition* at the end of this text.

The observer may or may not have had the same experiences that I have had but yet was moved by the artwork. What memories that caused me to produce the artwork, may or may not be the same memories that caused the observer to respond to the artwork. I have touched a memory in his brain in which he finds pleasure (reward). I have **empathized** with him and he with me. The question is. What is the brain doing with this shared empathy?

The key here is pleasure. Is there a **reward or pleasure system** in the brain? Is there a reinforcement area in the brain which captures essence? Is this where the warm fuzzy feeling comes from when I do a beautiful painting or write a beautiful poem (creativity)? Do any of these have survival value? (W)

Apparently, the desire for wanting something is in the limbic system and other areas (cortical and subcortical?) (W) depending upon the type of memory (long term). Emotions have survival value because they steer us away from danger (fear) and point us toward reward (pleasure). What I get as a warm fuzzy feeling for doing something well could be interpreted as satisfaction which results from elevated/inhibited neurotransmitter(s) levels. These elevated/inhibited levels of neurotransmitter(s) are a point of connection too (dopamine, serotonin etc.).

The key is that at the point of connection neither the artist nor the observer is probably experiencing the same memories causing pleasure in producing the artwork. Yet, both are experiencing connection even though they are usually separated in space and time. Are we talking about mirror neurons?

How does essence come in to this? Essence captures the essential elements of something without spelling out all of the particulars. It is a simplification process which the mind finds desirable. It is easier to grasp the meaning of something that is simple as opposed to that which is complex. Is simplicity beauty? It can be (# 160 *Flamenco Dancers*) but so can complexity (# 155 *Moondoggy*).

What is the Draw? It is a pull that draws you to do something. It has to do with incubation of an idea until you are ready to carry it out. The draw results from memories of life's experiences > incubation > maturation of readiness > the final pull to action. To take an artwork in your brain/mind and make it happen.

PATTERN RECOGNITION AND ESSENCE

(10/7/2020)

I was a hunter and particularly, a frog hunter when I was young (1954-1958). My friend liked fried frog legs to eat. He said that they tasted like chicken. I never ate any. I just hunted them.

Frog hunting is usually done in swamps which provide a good background of green, brown, and tan for the frogs which are similar in color. They croak but it is hard to pinpoint where they are based on the sound they make, so you have to train your brain to pull them out from the background by shape(line) and pattern. You have to know the essence of the frog to do this. This is called pattern recognition by me. Are mirror neurons = pattern recognition?

Pattern recognition is used in machine learning in artificial intelligence to screen big data for information hidden in the data which could not be readily seen by a person and to gather actionable intelligence on which a decision can be made.

Essence captures the essential elements of something without spelling out all of the particulars. This is what I did with the frogs. It is a simplification process which the mind finds desirable. Take artworks as an example. The artist produces his artwork. All artworks (paintings, wood burnings, poems) have any number of unspecified strongpoints when they are finished (Static)-the artist is done.

What are strongpoints? (associations) They can be anything having to do with design like color, shape, soft edges, complementary colors, etc. The observer matches his strongpoints with the artworks strongpoints without necessarily knowing it and finds that appealing and loves that, wants that, has to have that. The observer's strongpoints can be separated in time and space from the artist.

Where do the matching strongpoints come from? They can come from anywhere in the observer's experiences which produces memories and thoughts. His brain (mind) sees the match and he loves that, wants that, has to have that. This explains how the memories of the artist and the observer connect even though they are separated in space and time. It also explains the strength of attraction. The more strongpoints that match up in the observer with the artwork the stronger the attraction. Perhaps mirror neurons are simply matching strongpoints like matching strongpoints in an artwork!

How does this work? Let's say we have an observer sitting on a park bench years ago on a beautiful day and a pretty young lady in a beautiful orange dress walks by and smiles at him. This is a memory he cannot forget but may remain dormant until stimulated.

Fast forward to the present and he is at an auction and # 254 *Road Side Poppy* is up for sale. One of its strongpoints is a beautiful orange flower.

The observer immediately matches his strongpoint (**orange-its intensity enhanced by her smile and – contrast with the background?**) with the orange of the flower, the **size** of the painting and locks on and is attracted to it. Maybe he has fond memories of **flowers** of that **shape** each of which may have come from past experiences. He likes **complementary colors** for he uses it in his choice of dress. He may or may not know why. He may even like orange cream-cycles. Now, he has five strongpoints locked on bringing on a much stronger attraction to that work of art such that he loves that, wants that, has to have that and acts out and buys that painting (makes it happen)! The more strongpoints that he has based on fond memories the stronger the attraction.

His brain locks on spontaneously without him being aware of why and he loves it. This is pattern recognition frog hunting style. There was a time in science when it was fashionable to say, "If you can't put a number on it, it is worthless." My experiments refute that saying. I don't know what results the Mind Brain Institute finally got when they tried to put a number on "What is Beauty?" I will bet that the results were ambiguous at best. I am writing about pattern recognition which is qualitative (subjective) vs pattern recognition which is mathematical (objective).

Let's look at another color. Yellow! A song by Coldplay and attributed to Chris Martin and the band. Some say he wrote it for his wife Gwyneth Paltrow or a long-lost love. He said that it just **popped** into his head when he was taking a break and looking at the stars and that he and the band finished it off that night. It doesn't matter how it originated. It only matters that it was produced. He had to express that and he did. It is a beautiful finished off song. Some strongpoints are the color yellow, love, melody, and lyrics.

I come along and he is writing about a color I love (bright, cheerful, out-going) the lyrics tell a story about the author, his emotions bringing out my emotions. The vision of stars on a clear night. I bond strongly 4/4 points. I love it, have to hear it!

Does this imply a relationship between the auditory and visual cortex?

Moving on! The Poem, *"Tickle me with Feathers."* It is finished.

It **popped** into my head too. It is funny, sentimental, and sincere. The three stanzas are its strongpoints. I love it, I had to have that, and most importantly, I made it happen!

Now, let's see how a painting becomes a wood burning because of pattern recognition. This is a painting of the steam ship *Delmarva* (# 75) which was one of many ferries running between Cape Charles, VA and Little Creek, VA. I went on these ferries many times since I lived there during WWII. Here is a wood burning of it. I can still smell the Bay, the roll of the ship, the look of the waves trailing off of the stern, the smell of the smoke. I loved these ferries and I had to make them happen for others to see and love too. I painted them and made wood burnings (# 235) of them. The above strongpoints make me love these two artworks when I look at them. It explains how I can come back later and still enjoy a painting as both the artist and the observer.

When you see or hear something that attracts you, ask the questions how and why were you attracted to it and think about your life's experiences and memories that might have caused that attraction. Example: Love at first sight!

A LAST THOUGHT

Why have I done all of this? This has never been work to me. The only answer I can come up with is that "I was just having fun!" I just made the challenges into adventures which made life fun. And that fun carried me and gave me relief from the stresses of raising a family and work life. If you are drawn to do something (pulled), it is fun to do it and it results in "a work." If you are pushing (driven) to do something (berry picking), it *is* work and can be exhausting. I have been in both places (Before and after 1980). I may love the subject biology (a lifetime love of mine) but not the conditions of getting there. See my poem: *The Demon.*

QUESTIONS?

1. Can you put a number on beauty?

My results indicate that you can't. It is too individual.

2. Are (Is) stimuli to the senses causing the mind to recognize patterns (**pattern recognition**) and retrieve **memories/thoughts and their associations** stored as **mirror neurons** in the brain the same?

My results indicate that the brain stores stimuli and their associations through the senses as memory. The mind recognizes (pattern recognition) the stimuli and their associations stored in the brain (visual cortex?) and recalls it as a memory with associations stored and distributed along multiple(?)pathways (strongpoints). These memories and pathways are assembled as mirror neurons which can be detected with fMRI.

Your view of things is based on your lifetime of experiences stored in your brain giving you a unique view of the world (Qualia) expressed by your mind/personality. Hence, the observer does not need a signature to see my artwork sometimes as a "Stringfellow."

3. Are 2 above the essence of "My mind's eye...?"

I believe it is but how does visual cortex fit? The primary visual cortex is good at pattern recognition but does it explain how it recognizes subjective pattern recognition which I do? Yes! It's natural to me. Are the visual and auditory cortexes linked? I am an outdoorsman and a birder. Think of a bird, Barred Owl (visual), and its song "Who, Who, de Wooo!" (auditory!)

4. What controls all of this?

The cerebrum takes in data from the senses and makes sense of it (interprets it). It controls awareness and actions (voluntary activities). It plans, thinks, and judges and organizes speech and information. It carries out the "higher mental functions." *Wikipedia*. I can't say it better.

I view memories/thoughts as patterns on the plane of a map rolled out (Can this be done digitally?). A stimulus activates the senses which causes the mind to recognize that pattern stored in the brain, the mind recalls it, and with the personality acts out and produces an artwork. It likes what it sees, finds pleasure in it, wants it, and makes it happen. It is personal and individual and is an expression of that person's mind/personality (Qualia). The *Collected Works of Frank Stringfellow* is similar in preserving and expressing my life. The end result is that it is an expression of the self. You don't even have to see the signature in extreme cases. Example(s): a Modigliani, a Van Gogh, a Stringfellow. See: # 13 and 38 under Icon 6.

The memories/thoughts are interconnected by pathways and associations which allow them to interact and form new thoughts (imagination?) in the cerebrum. The map is open ended in its ability to contain memories/thoughts limited by the functioning of the body. The brain is dependent upon the body for its physiological needs and if the body fails to meet these needs and dies, the brain dies.

EXPERIMENTAL DESIGN FOR BRAIN MAPPING USING MIRROR NEURONS AND PATTERN RECOGNITION

I find it hard to believe that I arrived at this point by taking it one step at a time. It never entered my mind when I started that I would propose mirror neurons and pattern recognition as a way to map the human brain.

I become facilitated after I finish a work of art. I have to go away from it for a while. Later my mind being refreshed, I look at it and I say, "That really is a good work." I get that same warm fuzzy feeling from the neurotransmitters that are operating. I am now the observer.

See: #s 18 and 38.

An observer who sees that work may respond to it (light up) and get that warm fuzzy feeling too. Are they communicating in the same part(s) of their brains?

How do we test this?

Set up two fMRI, one for the artist and the other for the observer. Let there be a time delay such that each person is in a *refreshed state*, start the machines and show an artwork they both love to each person. You want them to "light up."

Do the scans show activity in the same areas of the brain scan? If so, this would reinforce the idea that mirror neurons are involved in the connection process. They may have had similar experiences. If not, there are other memories in the one causing the observer to "light up" in a different area of the brain. They tell another story about the observer.

Ask the persons what memories did they remember which caused them to "light up." Is this a way to map the brain for memories using pattern recognition?

What do you do if they don't "light up"?

Punt!

THERESE IN THOUGHT

367. *Therese in Thought*. Oil on canvas. 9 X 12 inches. Does this resemble using pattern recognition and mirror neurons to map the human brain? See: *Where Neuroscience Meets Art. Pattern Recognition and Mirror Neurons, Implications for Mapping the Human Brain* under Icon 15. Also See: *Experimental Design for Brain Mapping using Mirror Neurons*. There are a lot of similarities.

368. Photograph of: *Therese in Thought*.

 # 366. Therese in Thought. (5/2/2022). Monochrome-Payne's Gray. 9 X 12 inches. I was always fascinated by Andrew Wyeth's painting of a women in a pensive mood. These two drawings and paintings actually paint Therese in thought but it goes farther. It actually paints what she was thinking about. I actually painted a thought! She told me, "I was thinking who would be my bridesmaid when we got married" Her bride's maid was her sister, Kathleen Dunleavy. This confirms my pursuit of doing brain research.

What happened? She (her mind) recalled memories of Kathleen and her other sisters stored in her brain from a lifetime of living and inputs from her senses, acted out with her personality and chose her to be her bridesmaid in lieu of Mary and Patty, her other sisters. She was closer to Kathleen.

ABSTRACTION

How does abstraction fit into my concept of beauty?

I love Jackson Pollack's drip painting called *Lavender*. As an observer, it brings back memories of lavender which blooms in the fall on the marsh. I remember it well on our farm (Taylor Farm in Northampton Co, VA). It is subtle and beautiful (gossamer) much like he portrays in his painting. However, I wonder if he recalled actual lavender when he painted it (mirror neurons?). Maybe he just liked the color or the smell of the powder that little old ladies wear! Otherwise, I don't see much beauty in his drip paintings. Maybe, it popped into his head like the song *Yellow* by Chris Martin and the band.

I don't like Mondrian's paintings which my wife loves. She is mathematical (left brain) and I am both (scientist and artist). It looks to me that his abstract art is based on mathematical geometric thinking which I don't find beautiful. However, I remember a fellow student who was taking analytic geometry and he referred to it as beautiful. I guess he was describing the beauty of generating a geometric shape by using an equation. I believe I can understand that.

Abstract art then is based on the juxtaposition of color, shape, space and position relative to one another. It appears to be more of a logical process to provoke an emotion than that based on memory and feelings?

Complementary colors come to mind: blue/red = violet/; red/yellow; = orange; blue/yellow = green. These colors bring out (speak to one another) one another. Other than contrast, I can think of no other reason for these to be complementary.

The exception to this is gray/green which complements each other and has survival value (camouflage). Think of Mother Nature. Camouflage is the opposite of contrast.

Similar to complimentary colors is that artworks "speak" to one another. That is why placement is so important in a museum.

It is pushing to bring together combinations of shape, color, position and space pleasing to the "mind's eye" of the observer. That is why paintings are labeled "Untitled" because the artist has left it up to the observer to be attracted to it. What attracts the observer to it? It looks to me that the artist did not know why he did the artwork in the first place except that he knows that certain combinations will provoke an emotion! There is nothing wrong with that. Both the artist and the observer may have different reasons for doing/wanting the artwork.

I have been experimenting with abstracts. See: # 255-258. I used *Ammo lite (Ammolite-*proper spelling) to give meaning to color mixing. See: # 381-392. (3/2/2023).

FRANKIE'S BRAIN

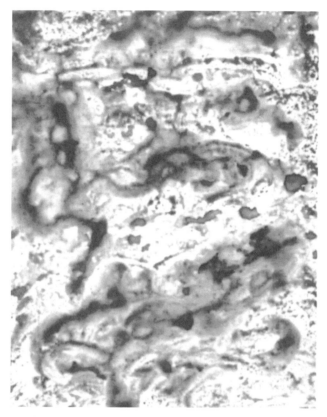

135. *# 389. Chaos. and (FRANKIE'S BRAIN) (2/11/2023).* Acrylic on canvas. 14 X 18 inches.

FRANKIE'S BRAIN

"This Must Be the Place."

Sis always said of me,
"You are the most discontented person
I have ever met."

She was right!
I have never had a moments peace in my life.

We were on our way
to Alaska through Vancouver airport
And
I found a little alcove where

it was warm, isolated, and I sat there dozing.

I had such a sense of warmth, calm, and peace.

This was "the place"

in the midst of a busy international airport.

A place of solitude, warmth, and contentment in time and space.

I knew what the poet meant when he said,

"This must be the place."

It was all inside of him.

You can try to see where you are going.

You can see where you have been,

But

Can you be (see) where you are now?

Is discontented (organized confusion) the same thing as chaos described in Dr. Schultz's book, *"The Mind of the Artist?"* See Books I found Useful. Is this the jumping off point where I go to find my "Happy Place?"

Refer to paintings 135. # 389, (Frankie's Brain 1) and 133 # 383, (Frankie's Brain 2). # 389 refers to the chaos that is described in Dr. Schultz's book *The Mind of the Artist.* # 383 shows a sense of calm and satisfaction I experience when I am being creative. Each one is never the same.

These thoughts summarize the theme of this paper, that I am pulled to do something creative to disrupt the chaos going on in my brain and find my "happy place." I use # 383 as an example but any of the creative artworks could serve as well.

The feelings I get when I like what I see, act out and do them, am satisfied in reading and doing them again are a result of neurotransmitters (dopamine and serotonin) operating. Apparently, these involve the reward system in the brain. There is then a physical and a chemical basis neurologically for these paintings. This is similar to what I describe in *"Therese in Thought."* #'s 366-368. (W)

REFLECTIONS OF A CLEAR DAY ON BLUE ICE

132. *# 386. Reflections of a Clear Sky on Blue Ice. (2/5/2023).* 9 X 12 inches. Acrylic on canvas. I saw the idea for this painting as I was going around the lake. I tried to use # 385 and its speckled areas to make blue ice and allow the clear sky to show through. This is a study for # 387.

FRANKIE'S FAVORITE COLOR

141. *# 395. Frankie's Favorite Color (2). (4/25/2023).* Acrylic on canvas 14 X 18 inches. Orange.

These feelings are temporary and have to be refreshed periodically much like rotating artworks in a museum. It is a matter of facilitation. You get bored by seeing the same old thing and being creative is refreshing.

Frankie's Favorite Color reached out and produced this painting which he had to have. Frankie went to his "Happy Place!" by painting *Roadside Poppy.*

ROADSIDE POPPY

58. # 253. *Roadside Poppy* with author (age 76)

254. *Roadside Poppy 3. (11/16/2016).* Oil on canvas. 64 X 74 inches. See it 127 and 149. I have decided to do some of my paintings large. It will make them more effective.

Is this the basis for art therapy, painting, crocheting, sewing, writing (poetry) etc:?

SOCIAL CONSCIOUSNESS

In 2007 I said that I thought that self-awareness (a sense of self) would be one of the great discoveries in systems biology (SRu 866-640 page 186). I am modifying that now to self-awareness/consciousness (S/C). Most sources seem to equate them as synonymous with the mind. I believe that they are integrated as the mind but consciousness is something much bigger than the self. It reaches out. It seeks. My concept of systems biology is holistic.

Awareness is the brain recognizing and integrating sensory input from all sensations at one given point in time and that this is changing from moment to moment. It is the brain's way of assessing sensory input in order to adapt to a constantly changing internal and external environment. It is an awareness of self and culminates in survival of the fittest. You eat what you kill!

There are various degrees of awareness. *Consciousness is awareness of being beyond self-awareness. It is open ended and is much bigger than the self.* It is able to see what is self-evident. It may even be the beginning for the search for reason (how or why) or for a Deity. It could also lead to survival of a species through natural selection of behavior with survival value beyond the individual (herd or pack protection).

If you want to see consciousness in action, look at the faces of those who have seen Nirvana (ex.: religious). There is a peace and calm written across their faces that transcends themselves. This is similar to what I call, "They light up." Is it possible that those who treat self-awareness = consciousness are incapable of experiencing consciousness, that they are too grounded in the immediate?

How does this fit in with art? (See WETA: *The Luminous Years,* 12/16/'10, the movie *Midnight in Paris, and Diaghilev and the Ballet Ruses, 1909-1929, When Art Danced with Music,* National Gallery of Art, 5/2-10/5, 2013). There was a time in post-impression art when great artists assembled in France near Paris for camaraderie and growth in a new direction to push the limits, probably in response to WW1. Gertrude Stein's Salon was a focal point for artists (Picasso, Gauguin, Chagall, and Matisse). There were poets (Apollinaire, Cocteau), writers (Hemingway), and musicians in addition to painters/sculptures. Many of these artists were multidimensional. Many new artistic movements arose during this time.

The one thing that they had in common was that they understood where they were going each in their own discipline but where the others were going too. They had an awareness in common (create new art work-push the limits) shaped by the times (input/culture/pull) to break new ground. They pushed one another to break new ground from which all of the new art works arose (modernism-post modernism). Did modernism-postmodernism have survival value? It already has and is. I said in the Foreword that I don't believe that there is only one way to do art. That modern art is what is being done now and this *Works* is my work of modern folk art.

Consciousness was the fertile (virgin) ground from which the creativity grew. It was man's curiosity (consciousness?) to create anew. They were aware from the input of the times (culture?) which impressed itself on the collective consciousness and which determined where the search would take them. Individually, they were multi-dimensional and the interaction between them blossomed into something *bigger than themselves.*

Other good examples where individuals went outside themselves and accomplished great things were Johns Hopkins and Charles Mayo whose visions carry on to this day. Others bought in and have carried the ball. And yes, they made money!

Perhaps, there are degrees of consciousness in humans from empathy to altruism to laying down one's life for one's friends. It is selfless whereas self- awareness is self-absorbed (selfish). Both have survival value. The first through natural selection leads to survival of the individual and the group and leads to cooperative behavior (herd or pack). The second leads to survival of the individual. We might be talking about some sort of social biology in which natural selection plays a bigger part in selection at a higher order in the former as opposed to natural selection of desirable traits or behavior patterns in individuals. Both can have survival value through natural selection since they both can or don't survive.

THE HUMAN CONDITION

I found myself doing artwork (photography, poetry, paintings, audios and wood burnings etc.) reflecting the human condition and me personally. What do I mean by the "human condition?"

It is everything experienced by you (Qualia) and what you do between birth and death.

It can be a deserted house in a field (# 121 or 122) and its slow decay or a fireplug and being ignored and lonely (# 123). The homeless, ill, and abandoned (#23-24). The effects of war (grief and death) and conflict on civilians and the military (# 50). A poem and art work about a person and their life (# 389). A completely out of control situation (#'s 351 and 354)! The influence of culture (# 2). What lies behind the red door, the uncertainties of life (# 119)? Just surviving (# 85). Having fun (# 130). Saying "Good-bye." (# 120) Love (#'s 18, 38, 184, 367). Memories of the past (# 207, 261, 265, 266, 366).

These are just a few that I selected out that have a deeper meaning. My view on that artwork is given in the text at that #. Nearly all of these #'s have cross-referencing to other #'s so that the expansion is enormous.

Most of the stories written and spoken (Icon 4) in *the Collected Works of Frank Stringfellow* have meaning for me. Especially the poems.

Pick out some artwork that has meaning for you as an observer. What memories do you associate with the artwork that caused you to like it and want it? And, there is so much more! (1/31/2023).

CLARIFICATION

Consciousness as we know it is something that comes from being alive (inherited) as opposed to being dead or inanimate matter. This may extend to every living creature.

We all experience consciousness **in our own way**. I use it to make an artwork?

My use of self-awareness and consciousness in this paper is limited to *What is Beauty?*

It is small part of a much larger view of consciousness I am only marginally qualified to comment on.

If the origin of the universe resulted from an intelligent energetic being who burst onto the scene with the "Big Bang", then it is not unreasonable for it to support and maintain the existence of matter, dark matter, dark energy based on Thomistic Philosophy of cause and effect. God is visible in his creations. See: # 45- *Cathedral of the Forest,* and my poem, A *Walk in the Forest.*

Perhaps consciousness is an emergent property of the universe. Maybe consciousness is **God**. See: Independent Lens, Aware, *Glimpses of Consciousness,* S 23Ep 12, 1 hr.: 25 min.

What happens when you die?

I am only experiencing a limited view of consciousness when I do artworks. There are those (religious) who experience it at a much higher level (Nirvana) and believe that their consciousness (soul) will be a part of that larger consciousness which we call God.

The following are essays I have written about Consciousness and related areas in my *Collected Works of Frank Stringfellow.*

THE BASIS FOR ALL OF THIS

This is the first poem that I wrote (1980) which led to over a thousand that I still have. I was working in the lab when I felt a compulsion to write. I got up and walked over to the center block bench and started writing. It only took a minute to do this.

I was given gifts which were stifled by the demon. These gifts were allowed to emerge when it went away allowing me to express these gifts. I started having fun. The expression of these gifts is through a pulling action. If I get something in my head, I am pulled to do it. It won't go away until I express it. See my essays *Full Circle* and *Consciousness, Awareness, and Art.*

I can't explain what happened here psychologically (mid-life crises?) but it was powerful and life-long. See my poem: *This Must Be the Place.*

The only other time I felt the compulsion to write like this was when I did *Glass.* See # 160.

Similar occurrences are *Tickle Me with Feathers* and *Pastels* (# 49-50). All of the woodburning were less spontaneous but equally as new. These were pop-ups.

THE DEMON

The Demon left today.
        Why he left or where he went I do not know.
I only know that he went away.

I knew Him well.
I knew Him long
        for forty years he rode along
                and drove and taunted me, the driven.
Is there no peace on earth, or in heaven?

Heaven is the loss of Him
        that black dog, that fear ridden malcontent
                that inner self, forward bent.
Striving for goals of early design
        taunted and threatened only to find
                discontent in tunnel vision and then,
Stagnation...
        A time when heaven met hell
                in the bayou of my soul.

The Demon left.
My burdens lifted.
My mind wiped clean of Demons past
        like guilt, fear, and loneliness.
I did not let Him go. He let me happen.

I never knew Him until now.
        And
I shall miss Him.

I never knew that He was me.

BOOKS AND ARTICLES I FOUND USEFUL
IN DOING THIS WORKS *(2/10/'14).*

* Books I read in 2018 and after.

Blackmore, S. and Troscianko, E.T. Consciousness an Introduction. Routledge: London and NY. 2018. 617 pp.*

Burns, Martha A. and Linda S. Hartsock. *Voices of the Chincoteague. Arcadia Publishing. 2007 p. 223.*

Carter, Rita. *The Human Brain Book.* NY: 2009 256 pp. I read this book around 2011-12 and it allowed me to develop some grasp of how the brain functions.

Du Toit, Johan, *How to Paint Portraits from Photographs.* Watson-Guptill Publications/ NY, 1992, p 128.

Gardner, Howard (1983), *Frames of Mind: The Theory of Multiple Intelligences.* Basic Books. *

Gazzaniga, Michael S. *The Consciousness Instinct Unraveling the Mystery of How the Brain Makes the Mind.* Farrar, Straus and Giroux, NY, 2018 pp. 274. *

Lewis, Jim. *Cape Charles A Railroad Town.* Xenophon Press. 2004, 2022. P. 307.

Livingstone, Margaret. *Vision and Art: The Biology of Seeing.* NY: Harry N. Abrams 2002. 208 pp. I read this book in 2014 and it allowed me to see better how the brain functions in painting, woodburning and in how I perceive nature.

Ramachandran, V.S. Phantoms in the Brain-Probing the Mysteries of the Human Mind, NY: William Morrow, 1998, p. 328.

Ramachandran, V. S. and William Hirstein. *The Science of Art.* The principles were published on the internet: *The Cognitive Science of Art: Ramachandran 10 Principles of Art.*

Ramachandran, V. S. *The Emerging Mind.* Profile Books. 2003. 208 pp. Updated version: A Brief Tour of Human Consciousness: From Impostor Poodles to Purple Numbers. 2004. 192 pp. P.I Press, NY.

Ross, Hugh. *The Creator and the Cosmos. How the Latest Scientific Discoveries Reveal God.* RTB Press Covina, CA. pp. 333, 2018. *

Schultz, William Todd. *The Mind of the Artist. Personality and the Drive to Create.* Oxford. 2022 200 pp. A useful book which confirmed many of the findings I found about myself as a scientist and an artist. I learned a lot from him he can learn a lot from me.*

Sheehy, Gail. *Passages: Predictable Crises of Adult Life.* NY: Bantam Books, 1977 564 pp. I read this book during the late '70's and early eighties. The ideas helped me to undergo personal development which allowed my gifts to emerge. See my poem: *The Demon.*

Zeki, S. A *Vision of the Brain.* Oxford: Oxford University Press. 1993.

BACKGROUND-WIKIPEDIA

Action Potential
Synapse- facilitation- refreshing
Electromagnetic theories of consciousness
Neurotransmitters
Nodes of Ranvier

Limbic System
Neuroanatomy of memory (long- and short-term memory)
Emotions

Cerebral cortex
Glutamate receptors

Mirror neurons- Pattern recognition. Coronavirus and the Wolf Tree
The Mind's Eye
Memories- Thoughts- Imagination
Primary visual cortex
Brain mapping
Imagination

Senses
Qualia
Theory of Mind. Empathy
Self-awareness-Consciousness-obsessive compulsive disorder (OCD)
Reward System
Five Factor Model (FFM)- Big five personality traits

fMRI (functional magnetic resonance imaging)
TMS (transcranial magnetic stimulation)

Modernism
Postmodernism
Gertrude Stein

Proof of concept
The Human Condition

This has always been an adventure and fun.

I received the Maryland Governor's Award for Leadership in Aging in 2011. The alternative title to this *Collected Works of Frank Stringfellow* is *Effects of Aging on Creativity and Expression.*

I encourage those who see value in what I have done to pursue their own ideas with it, to make it happen.

I HAVE SHOWN THE WAY!

BIOGRAPHY

My wife wrote this Bio. She said that the one I wrote was much too modest.

Dr. Frank Stringfellow, scientist, author, poet, painter, and storyteller, was born in Northampton County on the Eastern Shore of Virginia October 27, 1940. He received his early schooling in the public schools there (Atlantic High School, now Arcadia). Surrounded by creative, talented, and well-educated family members, Frank absorbed a love of nature, his fellow man, and storytelling that knew no bounds. His family's early roots and extensive web of kin filled him with a rich repertoire of life's stories. His grandfather, Dr. John Gates Goode, delivered over 5,000 babies including Frank! His great grandfather and namesake was the legendary Civil War Confederate scout, Frank Stringfellow.

Frank graduated from St. Louis University with a degree in biology in 1962. He attended the University of Missouri as an unclassified graduate student for the fall semester. In 1964, on a fellowship/internship he earned a Master of Arts degree in biology from Drake University. From there, he was offered an instructorship at the University of South Carolina where he completed his Ph.D. in biology in 1967. He obtained advanced training in pathology through the Veterans Administration and studied further in pathology at the Armed Forces Institute of Pathology.

He won a position as a research scientist at the world-renowned lab, The Animal Parasitology Institute, United States Department of Agriculture, Beltsville MD. Dr. Stringfellow worked for over thirty-one years as a research scientist and authored over forty scientific publications, most numerously in *The Journal of Parasitology.* He is listed in *American Men and Women of Science.*

Frank is now retired and enjoys family especially his three young granddaughters. He continues a lifelong passion for sharing his love of nature and expression as described in the Curriculum Vitae.

If you really want to see how I was raised, go to: *Growing up on Tilghman Island, Channel 22 Maryland Public Television, Wed. 4/24/"13, 12:30 a.m.* This hits it on the head!

Burns, Martha A. and Linda S. Hartsock. *Voices of the Chincoteague. Arcadia Publishing. 2007 p.223.*

Lewis, Jim. *Cape Charles A Railroad Town.* Xenophon Press.2004, 2022. p. 307.

Collected Works of Frank Stringfellow, Library of Congress Copyright Office, TXu- 2- 267- 678 (6 / 18 / 2020).

Frank Stringfellow. *The Treasure of Assateague Island.* 2005.

Frank Stringfellow. *Sandy and the Dancing Waves.* 2006. This children's book shows who Frankie is and why he did what he did in his life. Like Sandy, he didn't want to be like all of the other grains of sand on the beach!

CURRICULUM VITAE

Frank has authored over forty scientific publications. He has produced three literary pieces of which two are self-published

As a painter, Frank has produced landscapes, still-life, wildlife portraitures as well as family and urban scenes. He is an expressionist and colorist. His natural affinity with open skies and the flow of rural life is a result of his early upbringing on the Atlantic Ocean and the Chesapeake Bay of the Eastern Shore of Virginia. Frank has sketched scenic and thought-provoking paintings from Gettysburg, Pa. to the Inner Harbor of Baltimore.

Frank's woodburning of Eastern Shore scenes are especially coveted. *Oyster, Va.*, and fishing trawlers make up some of the best of this truly unique medium. Frank also enjoys building World War II historic paper/wood models.

Author Invitations:

Meet the Author, Book signing author at The Kite Koop Book Store, Chincoteague Island, VA summer, 2005 and summer 2006.

The Dan Gaffney Show, Invitee to WGMD Talk Radio Show, Rehoboth Beach DE, July 13, 2006.

Meet the Author, Book signing author at BrowseAbout Books, Rehoboth Beach, DE, July 15, 2006.

Donations:

Frank was invited to deposit CD's similar to this one with the following places: The Library of Virginia (9 /9 /'08, 3 /15/ '16, 7 / 2020), The Virginia Historical Society (9 /9 / '08, 1 / 2013, 3 /15/ '16, 7 / 2020), both in Richmond, Va., the Alexandria Public Library (9 /17/ '08) in Alexandria, Va., *Carlyle House* (8 /15/ 2015) in Alexandria, Va., the Accomack County Library (10 /10/ '08, 3/15/'16) in Accomack, Va., The Library of Congress where he deposited his *Collected Works of Frank Stringfellow* {Library of Congress:TXu-043-426 (7 /19/ '01), SRu 561-934 (4 /26/ '04), SRu 866-640 (8/30/'07), SRu1-046-548 (6/3/2011) }, SRu1-293- 075 (3 /16/2016) (and the Museum of the History of Culpeper (2 /3/' 2012).See # 28. He submitted a flash drive of his *Collected Works of Frank String fellow* to the Smithsonian Hirschhorn Museum of Art for future reference. They recognize it as a body of works, some national recognition [(See the letter from them under Hirschhorn and reply (8 /13/ '10)].

Frank was nominated by his wife for the MD Governor's Leadership in Aging Award for the Visual Arts. He received the award on Tuesday, May 10, 2011. The flash drive that he sent in to the Dept. of Aging with his *Collected Works* is on file (John Murphy).

Frank was nominated for a *National Heritage Fellowship for the National Endowment for the Arts* by Mary and Steve Daly on 9/24/'12. He did not win anything but they did contact the MD art advisory committee so that at least means that they knew that he exists, a form of national recognition.

A researcher can trace the evolution of this *Collected Works* by using the copyright dates.

Presentations and Talks:

The Museum for History of Culpeper, VA. 2 /4/ 2012 Star Exponent article.

Carlyle House. Alexandria, VA. 8/182015. A small talk.

Read Frank's letter to Emma at marking of graves ceremony sponsored by the United Daughters of the Confederacy. See photos of ceremony at ICON 13.

Memberships, Honors, and Consultant Activities:

The Society of the Sigma Xi: Honorary scientific organization for the promotion of research in science.

Superior Achievement and Service Awards:

Agricultural Research Service, U.S. Department of Agriculture.

American Society of Parasitologists: Presenter/Editor

Helminthological Society of Washington: Program Presenter/Editor

Southeastern Society for Parasitologists/Biologists: Assistant Chairman, Auditing Committee/Presenter

Elsevier Publishing Company: Consultant on foundation of international scientific journal of parasitology.

Maryland Governor's Award for Leadership in the Visual Arts (May 10, 2011), Invitation as Presenter, #'s: 2, 20, 76, 150, and 158.

Keynote speaker at The Museum of the History of Culpeper on the Anniversary of the Civil war in Culpeper. See # 28.

Special Honors and Recognition:

Frank received the MD Governor's Award for the Visual Arts on 5/10/'11. It was for his paintings and his children's books: *The Treasure of Assateague Island* and *Sandy and the Dancing Waves*. In addition, he received a certificate of special recognition from U.S. Senator Benjamin Cardin for service and achievement, as well as a certificate of excellence in the visual arts from U.S. Senator Barbara Mikulski. These are United States Senators so that is recognition at the national level as well as the state level. He was also presented with an Official Citation from the General Assembly of Maryland congratulating him for being selected to receive this award. They also presented him with a large poster showing *Assateague Lighthouse with Horses,* #20 which covered the entire cover of the program. He has also donated to and has his *Collected Works* in the Museum of the History of Culpeper.

Frank is recognized as a visual artist at the state and national level. See above. He was nationally known as a scientist. See his scientific publications and note # 2.

Maryland Senior Olympics, 1996-2005, member-participant and recipient of bronze, silver, and gold medals.

National Senior Games: Invited to participate in Senior Games in Houston, TX, June 16-2011, 100 meters free style. See letters under genealogy. I won my heat and placed 18th in the nation of 22. I did the 100-yard free style in 1.37.69 minutes. My qualifying time was 154 at Germantown. I did the Eastern Shore games in Salisbury at 143 so I kept getting better.

Wildflowers of Maryland, course taught by Dr. Stringfellow through Greenbelt's Community Arts Program. I identified about 200 plants in conjunction with this class and my own interests.

Presenter: Federal Duck Stamp Contest, 1996-2004.

Maryland Duck Stamp Contest, 1996-2004.

Maryland Black Bear Contest, Presenter three years.

Painter for Acme Clown Circus.

Maryland Society of Archeologists.

An Artist's Statement can be found under paintings.

Printed in the United States
by Baker & Taylor Publisher Services